Bridging the Skills Gap in Chemical Engineering

Held at the 2020 AIChE Annual Meeting

Online
16 – 20 November 2020

ISBN: 978-1-7138-2060-4

Printed from e-media with permission by:

Curran Associates, Inc.
57 Morehouse Lane
Red Hook, NY 12571

Some format issues inherent in the e-media version may also appear in this print version.

Printed with permission by Curran Associates, Inc. (2021)

AIChE
120 Wall Street, FL 23
New York, NY 10005-4020

Phone: (800) 242-4363
Fax: (203) 775-5177

www.aiche.org

Additional copies of this publication are available from:

Curran Associates, Inc.
57 Morehouse Lane
Red Hook, NY 12571 USA
Phone: 845-758-0400
Fax: 845-758-2633
Email: curran@proceedings.com
Web: www.proceedings.com

TABLE OF CONTENTS

612375 Leading and Managing Industry-Academia Partnerships and Collaborations Panel Discussion

Abstract:

Outline:

- Discussion on benefits resulting from and concerns hindering industry-academia collaborations
- New methods and long term changes needed to lead and manage industry- academia partnerships and collaborations
- Benefits to diverse sectors including R&D, manufacturing, economics, strategic planning etc. and increasing participation across all.
- Definition of research contracts, communication channels, international collaborations, project timelines and deliverables

Invited Panelists:

Barclay Satterfield: External Innovation Leader at Eastman Chemical Company who has experience in open innovation and industry/university research partnerships, polymer property research and environmental life cycle analysis. She has been involved in a $10M 6-yr research partnership between Eastman Chemicals and NC State University.

Kenneth Rueter: President and CEO - UCOR LLC, a past Management Award recipient has over forty years of experience. He has been involved in a highly successful collaboration with the University of Tennessee Nuclear Engineering Department. This collaboration has established a Nuclear Decommissioning minor at the University of Tennessee enabling students to meet the high demand for Nuclear Legacy Cleanup around the World.

Dr. Kiarash Vakhshouri: Young professional working with Google with a solid expertise in structural characterization of semiconductor materials, thin film coating and printing, rheology, lithography and designing and operating laboratory and pilot plant chemical processes. He has several well-cited publications and four US patents to his credit and has been engaged in a number of industry-academia collaborative research projects right since graduate school.

Dr. Phil Westmoreland: Has a PhD from MIT and has been a Professor of Chemical and Biomolecular Engineering with the North Carolina State University. He has over 35 years of academic research experience and has held a variety of Directorial positions during his illustrious career. He has also served as the AIChE President in 2013. His research interests include flame chemistry, theoretical and computational chemistry and polymer decomposition kinetics and thermochemistry. He has been instrumental to a number of industry-academia-government collaborative projects.

Dr. Quinta Warren: CEO at Energy Research Consulting, PE, PMP has experience in industry, academia as well as Government. She was working with ConocoPhillips, has been a director of professional education and also worked with the US DOE. She brings in a unique entrepreneurial skill-set into synergizing these areas.

Dr. Sarma Pisupati: Professor of Energy and Mineral Engineering/Chemical Engineering and Director of Center for Critical Minerals at Penn State. He has more than 35 years experience in academia and has prior industrial experience as well. He specializes in computational fluid dynamic modeling of combustors, coal/biomass gasification, and biomass based materials for emission reduction. He has driven several successful collaborative research projects and novel initiatives with the US DOE as well as industrial partners.

Samuel Tam:(Have reached out only this week and waiting to hear back) President and CEO of CAC Solutions LLC and has been with the US DOE for 9 years where he managed total annual R&D spending was over U.S. $100 MM. He also has experience as a China Chief representative while working with the US DOE. Over his forty years of R&D, engineering, business development, pilot plant design and operation, and commercialization of technologies related to climate change in US DOE and five other major US corporations, Samuel has gained immense experience on fostering collaborations.

601991 The Digital Transformation of the Chemicals Industry - and How to Prepare

Abstract:

Over the course of the next 20 years, the chemicals industry will see more change than it has in the last 50, in large part due to digital technologies. Digital transformation is affecting business processes from R&D, to manufacturing, to supply chain, to sales and go-to-market -- and the data from all these digital efforts is tying those diverse functions to optimize the chemical enterprise and create new business models and sources of value. This presentation looks at how the industry will change through 2040 and examines what the new sources of competitive advantage will be for both companies and individuals.

613114 Driving the Value from Digitalization

Abstract:

Many opportunities exist to utilize the technologies of Industry 4.0 and digitalization within the process industries, to improve reliability, productivity, profitability, and safety. However, organizations may become enamored with the new tool or capability, looking for problems to fit the technology rather than focusing on the true business needs. Only by being technology agnostic and using a strategy that places balance sheet impact and long term competitive advantage as selection criteria to the myriad of solutions available in the market will an organization focus on the proper challenges. This presentation will examine potential pitfalls to technology selection and implementation.

613028 Combining AI with Chemical Engineering First Principles Will Take Process Performance, Safety and Sustainability to New Levels

Abstract:

Digitalization is now in the spotlight in terms of creating improved agility, resilience, and ability to operate process industry assets remotely and safely.

Sensors, connectivity, and cloud computing are making process data more available for use in optimization and prediction. High performance, edge computing, and new algorithms are enabling placing process models online for monitoring, optimization, and operational advice.
Most process and chemical engineers in industry have observed the influx of data science and data scientists into their organizations. This session will cover an area called "hybrid modeling." This combines and embeds AI and machine learning within first-principles rigorous models. You may think of this as putting chemical engineering guardrails around machine learning. Rigorous, first principles models, brings to bear decades of chemical engineering knowledge. Machine learning and AI brings to bear ability to rapidly analyze, and progressively learn from empirical operating data. When combined, it is now possible to achieve a new level of modeling that will further leverage both types of analysis. We have been validating this breakthrough combination of AI and chemical engineering with over 100 companies in industry, and will cover several high value and immediately practical use cases during this session.

601992 Designing Materials and Processes in the Digital World

Abstract:

3M has been known as a materials innovation company for many years with thousands of new products launched annually, many of those enabling and empowering everyday products. This innovation model is based on leveraging deep technologies into novel combinations to develop product platforms.

This talk will address the challenges and progress in transforming this culture and enabling it with digital tools created internally and with outside partners to accelerate the speed of materials innovation dramatically.

NOTES

NOTES

NOTES

NOTES

NOTES

NOTES

AUTHOR INDEX